A GUIDE TO THE WORLD OF THE

# JELLYFISH

From far out in space, the planet Earth sparkles bright blue. It's a jewel-like planet, made incomparably beautiful by its most remarkable feature—water. Earth is covered by oceans so wide and deep they dwarf its continents. Diving into this water world, you enter the realm of the jellyfish. Pulsating gracefully, arms trailing like delicate lace, yet powerfully armed for foe or prey, these creatures are one with their surroundings. Alien as they seem to us, they are at home in this otherworld.

Let this guide take you to a world beyond our own—the World of the Jellyfish.

# THE WORLD OF THE JELLY

A purple-striped jelly, *Pelagia*, resembles a Tiffany lamp in design and size.

Earth might better be called Oceanus. Fully two-thirds of the globe is covered by water, and from the lighted surface to its dark depths is a journey of more than two miles. As you dive into the ocean, you enter a habitat larger than all of Earth's deserts, forests, prairies and lakes combined. So vast that no one has explored it thoroughly, this place is populated by beings that are equally mysterious—the jellyfishes, or jellies.

Life on Earth began in the ocean, and in many ways the living is still easy here. The ocean's currents provide free transportation, and the water's alive with shifting clouds of food. Only in such a fluid world could something as delicate as a jelly survive. Water holds the jelly in a gentle embrace, so it doesn't need muscles or bones for support. A jelly is 97 percent water, and has only the thinnest of skins. Washed ashore, it would start to dry out

Oceans cover more than two-thirds of Earth's surface.

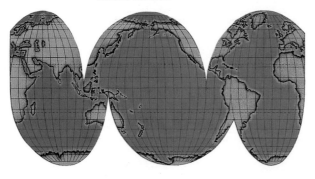

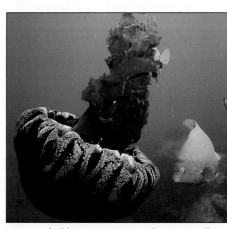

Two garibaldis investigate a *Chrysaora* jelly.

## THE WORLD OF THE JELLY

immediately. But living in water—in its element—a jelly needs little to protect it from its environment.

Simple in design, fragile of build, jellies have few of the complex features many animals use to survive. Yet for 650 million years, they've lived and prospered on this watery planet.

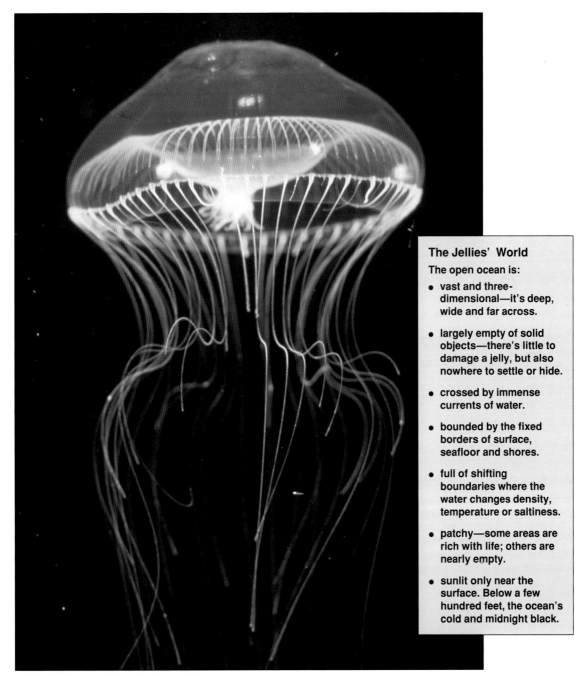

**The Jellies' World**
The open ocean is:

- vast and three-dimensional—it's deep, wide and far across.

- largely empty of solid objects—there's little to damage a jelly, but also nowhere to settle or hide.

- crossed by immense currents of water.

- bounded by the fixed borders of surface, seafloor and shores.

- full of shifting boundaries where the water changes density, temperature or saltiness.

- patchy—some areas are rich with life; others are nearly empty.

- sunlit only near the surface. Below a few hundred feet, the ocean's cold and midnight black.

Like many jellies, this crystal jelly, *Aequorea*, is almost transparent. It's a useful camouflage in the open ocean where there's nowhere to hide.

# THE WORLD OF THE JELLY

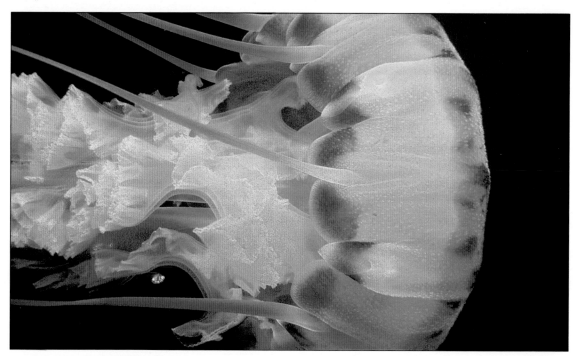

A purple-striped jelly, *Pelagia*, pulses along. Ruffled mouth-arms surround the jelly's mouth and stream behind with its yard-long tentacles. The dimples on the jelly's bell hide its sensors for gravity and light.

In the society of the sea, jellies practice the trade of drifting: they belong to the plankton, plants and animals that live by riding the ocean's currents. Aimless as it may seem, it's a common way of making a living in the sea.

Hitching a ride on the currents, a jelly can travel far and wide with little expense of energy. And the prey that jellies eat are also drifters, swept up in the same currents the jellies travel. By going with the flow, jellies end up swimming in rich drifts of food.

To be carried by the currents, drifters have to travel light. Jellies are made largely of a jelly-like stuff. It gives them size and substance, but lets them be almost weightless in water. And this "jelly" is transparent, so the animals can almost disappear—a useful trick here where there's no place to hide.

There really isn't much to a jelly. It's not a fish; it has no brain, no heart, no real eyes. A network of nerve cells helps it move and react to food or danger. Simple sensors around the rim of the bell let the jelly know whether it's headed up or down, into light or away from it.

But with this basic equipment, jellies manage to defend themselves from danger, make daily and seasonal journeys, stay together and occupy all the oceans of the world. These seeming blobs of jelly hold critical positions in the ocean's food web. They are, by any definition, a supremely successful group. They have answered the complex challenges of their world with elegant simplicity.

A jelly's mouth-arms shelter a young fish.

**Right:** The tentacles of this young sea nettle, *Chrysaora*, stretch to three feet while it hunts.

*Human subtlety...will never devise an invention more beautiful, more simple or more direct than does nature, because in her inventions nothing is lacking, and nothing is superfluous.*

Leonardo da Vinci

# CATCHING A MEAL

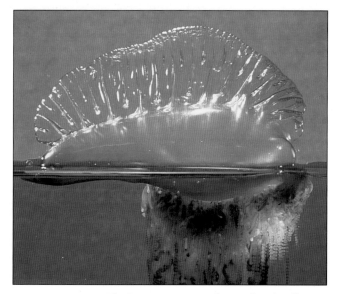

The Portuguese man-of-war, *Physalia*, floats at the surface (above) and catches meals with a stinging curtain of tentacles (below and lower right). From the jelly's float, the tentacles can stretch down 100 feet or more. They're armed with countless stinging cells, so don't swim too close—even a single tentacle swipe can leave a burning welt.

### Eating on the go

Jellies dine as they drift. They catch and eat the smaller plankton of the sea: larval fishes and invertebrates, little shrimplike animals, other jellies. Jellies don't have the senses or the speed to actively hunt their prey. But as they swim or drift along, they spread a live-wire net of tentacles. When food blunders in, they react with a quick, deadly reflex to catch it.

### Dining customs

A jelly's tentacles, and often other parts, are studded with stinging cells. When the tentacles brush against an animal, thousands of these cells explode, launching barbed stingers into the victim. The stingers, called nematocysts, inject paralyzing poison through long, hollow tubes. Other nematocysts entangle prey instead of stinging it.

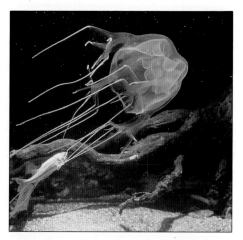

**Above:** This is the ocean's most dangerous jelly: the Australian sea wasp, *Chironex*. Beware the sea wasp—its sting can kill a person in minutes.

## CATCHING A MEAL

### Traveler's advisory

While most jellies are harmless to humans, a few are dangerous or even deadly. The stinging ability that makes them effective predators is also useful for defense. If you visit the jellies' world, watch out for trailing tentacles and try to avoid them. Jellies don't sting people on purpose, but the results can be painful nonetheless.

A sea nettle, *Chrysaora*, shows the stinging tentacles and mouth-arms that inspired its name.

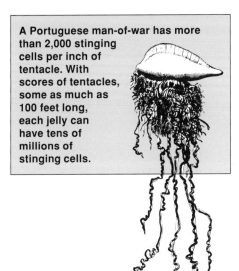

**A Portuguese man-of-war has more than 2,000 stinging cells per inch of tentacle. With scores of tentacles, some as much as 100 feet long, each jelly can have tens of millions of stinging cells.**

**Next page:** These *Cephea* jellies are the size of punch bowls, but like most jellies, they're harmless to humans.

**Here's how a jelly stings:**
(This drawing is 11,000 times larger than life.)

1. The stinger is triggered when it touches the victim's skin.
2. The cell bursts open, launching the stinger into the skin.
3. Anchored by a barb, the stinger shoots a long tube into the wound and injects a poison through it.

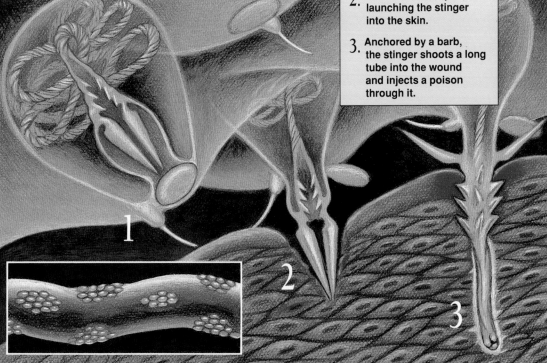

**Inset.** This section of tentacle (400 times life size) shows how the stingers are arranged in clusters.

# GETTING AROUND

This spotted *Mastigias* jelly swims alone, but jellies usually travel in large groups.

Jellies have two ways of getting around their world. Each method of moving serves a different purpose.

## Catching the currents

For traveling long distances, jellies hitch a ride on ocean currents. The currents are a mass-transit system—so many jellies ride the seaways that sometimes the water is packed with them as far as the eye can see.

Currents are cheap transportation, but riders have to go where the routes carry them. The jellies you see washed up on the beach or tumbling in the surf have unknowingly caught a death-bound current.

**Left:** Each day, masses of stingless *Mastigias* jellies travel half a mile across the saltwater lake where they live, then back again. This daily swim keeps them in the sun—important because these jellies are farmers. Each grows a crop of algae in its body, and from it harvests most of its food.

## GETTING AROUND

**Being self-propelled**
Jellies aren't totally at the mercy of the currents; they can also swim on their own power. Their graceful pulsing is a form of jet propulsion—each pulse sends a stream of water jetting out from the jelly's bell, propelling the animal in the opposite direction.

Jellies swim on shorter trips, giving direction to their drifting. Many make daily journeys up and down in the water. They swim toward the surface at night to feed, then back down when daylight comes to hide in darker depths. Other jellies make trips across the water, traveling to stay in the sun or to gather for mating.

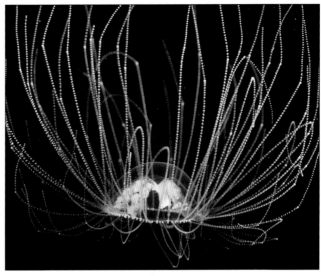

A *Gonionemus* jelly drifts slowly downward, tentacles outspread to catch small prey.

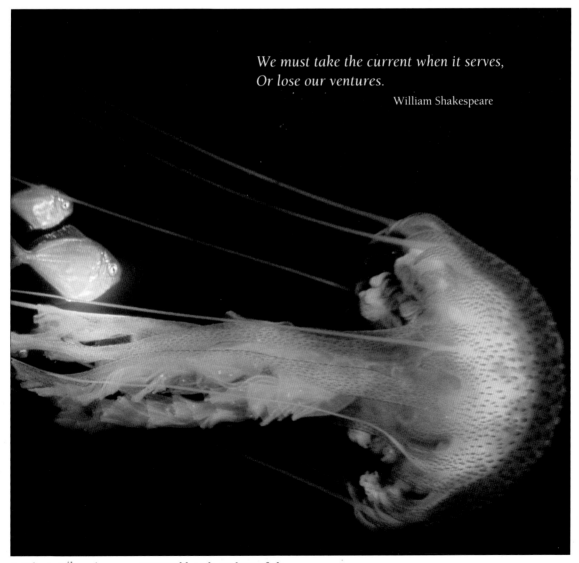

*We must take the current when it serves,*
*Or lose our ventures.*
                                William Shakespeare

A *Pelagia* jelly swims, accompanied by silvery butterfish

# THE PRIVATE LIVES OF JELLIES

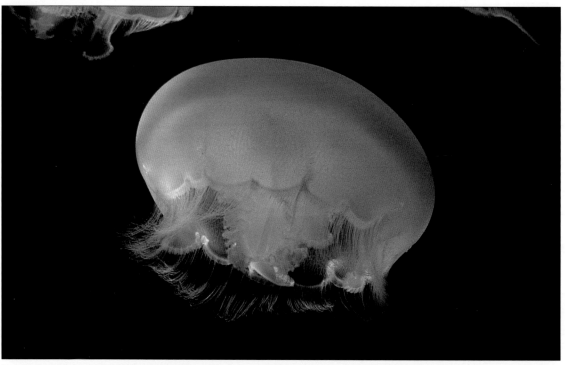

In spring and summer, adult moon jellies, *Aurelia*, show up in many oceans of the world. They disappear in tatters with the winter waves, but leave behind a crop of tiny polyps to produce the next generation.

Nothing about jellies is more alien to our experience than their complex and mysterious private lives.

## The visible adults

The familiar umbrella-shaped jellies you see pulsing through the water are only one part of the jelly life cycle. They're the adults, each either male or female, and they reproduce in a familiar way: by combining their eggs and sperm to make young.

## The hidden young

These young, called larvae, begin the most mysterious part of a jelly's life. Attaching to a solid surface, each grows to resemble a tiny flower. This stage, called a polyp, is as different from the adult jelly as a caterpillar is from a butterfly. They're so different, so small and elusive, that the polyps of some species have yet to be discovered.

In hidden caverns and under rocky ledges, polyps perform their own kind of reproduction—not with eggs and sperm, but by cloning themselves. First they produce identical new polyps. Then they begin to form free-swimming jellies, a process as strange as if a single caterpillar could divide itself into dozens of butterflies. A polyp can live for years, budding a new crop of jellies each spring, but adult jellies rarely survive the winter's storms.

During the first part of a jelly's life, it looks nothing like it will when it's an adult.

# THE PRIVATE LIVES OF JELLIES

**The Life of a Moon Jelly**

1. The female moon jelly's eggs are fertilized by sperm from a male.
2. When the eggs hatch, the larvae swim off in search of solid surfaces.
3. Each larva settles and grows into a polyp.
4. The polyp splits itself into many flat segments.
5. The segments peel off one at a time, and each swims away as a flat young jelly called an ephyra.
6. The ephyra grows into an adult moon jelly.

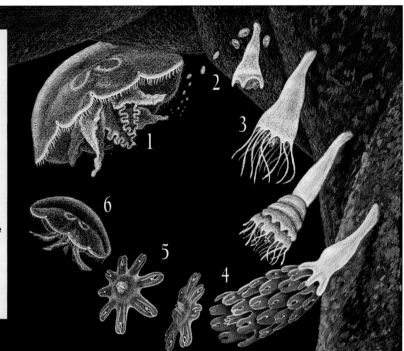

A female moon jelly holds clusters of white eggs on the lacy mouth-arms under her bell.

Jellies can clone themselves. During the polyp stage, each polyp produces dozens of identical polyps, budding them from its stalk like a strawberry plant buds runners. A few polyps can soon blanket an area with thousands of their clones.

Moon jellies have horseshoe-shaped sex organs in which their eggs or sperm are formed.

# AN ALBUM OF JELLIES

Earth's oceans are filled with a surprising variety of jellies. They live in every sea, and even in some freshwater ponds. They range from the equator to the poles and from the surface to the depths.

The variety in their forms and habits is a reflection of their varied lives: some from the dark midwater are highly colored, while many from sunlit waters are nearly transparent; some have floats for sailing along the surface, while others have stalks for attaching to the bottom; some have large, stout tentacles for catching good-sized prey, while others have a fine fringe for snagging tiny food.

Siphonophores like this *Physophora* (above) are some of the most bizarre jellies of all. They aren't even single animals—each is a colony made of many smaller members living and working together as a coordinated unit.

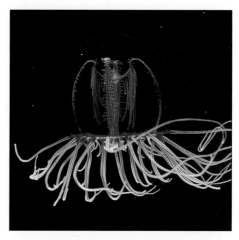

Many jellies live near the shallow seafloor. The bell jelly, *Polyorchis* (above), swims in eelgrass beds and often rests on the bottom. The stalked jelly, *Haliclystus* (below), attaches to a rock by its stalk.

> **With its tentacles fully stretched, the Arctic lion's mane jelly is probably the longest animal on Earth—longer than a 100-foot blue whale. The bell of this jelly can be up to six feet across.**

These transparent beauties are the size of small marbles.

## AN ALBUM OF JELLIES

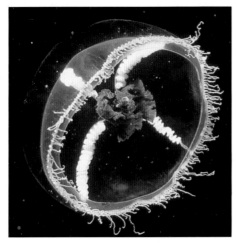

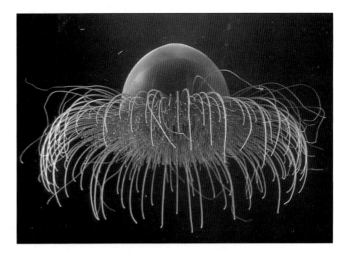

Jellies from the midwater and the deep sea are often colored dark purple or red. Some species grow to four feet across, but these range from grape- to grapefruit-sized. Clockwise from above, they are: *Tima, Benthocodon, Periphylla, Aeginura* and *Aglantha*.

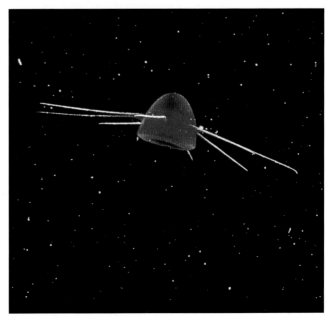

# THE GREATER COMMUNITY

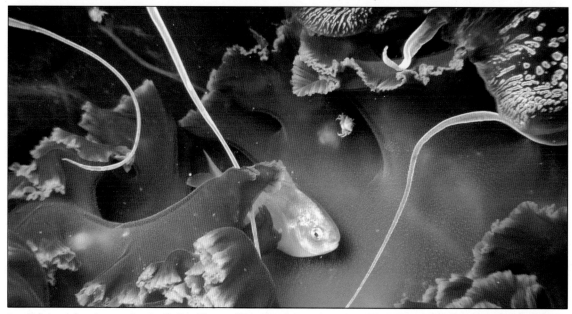

Small fishes often live under the bells of large jellies. The fish are protected by a fence of tentacles, and may also feed on the jelly's leftovers. No one's sure how they avoid being stung—they may just learn to dodge the tentacles.

If you visit the world of the jellies, you'll discover a whole wide community of which jellies are only a part. Watch them and you'll begin to sense the network of connections among all the plants and animals in the sea.

### Fishing for a living

Jellies make their living by fishing for smaller drifters. A large fleet of fishing jellies can nearly clear an area of prey. In some parts of the sea, jellies outnumber all other predators.

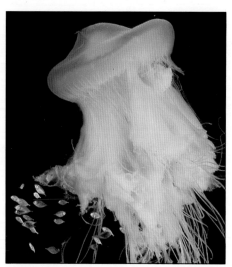

A giant purple jelly, *Drymonema*, shelters small fish.

Some sea turtles have a taste for jellies.

### Jelly enemies

Jellies sometimes fall victim to other animals. Sea turtles, ocean sunfish and blue rockfish all eat jellies, seemingly unaffected by the stings they must receive. Ocean-going snails nibble on them, and even seabirds have been known to dine on jelly swarms.

### Community relations

Jellies can be community resources. In a world without many solid objects, jellies offer shelter to small animals. Young fishes hide in their tentacles, and small invertebrates hitch a ride. Sometimes the relationships work two ways: jellies that shelter algae within their cells get food from photosynthesis in return.

**Right:** A *Lychnorhiza* jelly